Disclaimer

The publisher of this book is by no way associated with the National Institute of Standards and Technology (NIST). The NIST did not publish this book. It was published by 50 page publications under the public domain license.

50 Page Publications.

Book Title: Guide on Data Entity (Naming Conventions)

Book Author: J J. Newton;

Book Abstract: A coherent set of naming conventions for data entities is crucial to the central management of data. Name content and format must be designed to maximize the information content and relationship to the logical structure of the data. This report discusses the development and administration of naming conventions for data entities. The Information Resource Dictionary System (IRDS) meta-name schema provides a framework for name development. This report describes a methodology for deriving a relationship between an entity's dictionary names and details a method for structuring the format and content of entity names which maximizes opportunities for analysis and sharing of data.

Citation: NIST SP - 500-149

Keyword: data administration; database; data dictionary system; data engineering; data entity naming conventions; data management data standards; IRDS

PB88-123799

Computer Science and Technology

NBS Special Publication 500-149

Guide on Data Entity Naming Conventions

Judith J. Newton

Information Systems Engineering Division
Institute for Computer Sciences and Technology
National Bureau of Standards
Gaithersburg, Maryland 20899

October 1987

U.S. DEPARTMENT OF COMMERCE
Clarence J. Brown, Acting Secretary

National Bureau of Standards
Ernest Ambler, Director

REPRODUCED BY
U.S. DEPARTMENT OF COMMERCE
NATIONAL TECHNICAL
INFORMATION SERVICE
SPRINGFIELD, VA 22161

Reports on Computer Science and Technology

The National Bureau of Standards has a special responsibility within the Federal Government for computer science and technology activities. The programs of the NBS Institute for Computer Sciences and Technology are designed to provide ADP standards, guidelines, and technical advisory services to improve the effectiveness of computer utilization in the Federal sector, and to perform appropriate research and development efforts as foundation for such activities and programs. This publication series will report these NBS efforts to the Federal computer community as well as to interested specialists in the academic and private sectors. Those wishing to receive notices of publications in this series should complete and return the form at the end of this publication.

Library of Congress Catalog Card Number: 87-619867
National Bureau of Standards Special Publication 500-149
Natl. Bur. Stand. (U.S.), Spec. Publ. 500-149, 57 pages (Oct. 1987)
CODEN: XNBSAV

U.S. GOVERNMENT PRINTING OFFICE
WASHINGTON: 1987

NBS-114A (REV. 2-8C)			
U.S. DEPT. OF COMM. **BIBLIOGRAPHIC DATA** **SHEET** (See instructions)	1. PUBLICATION OR REPORT NO. NBS/SP-500/149	2. Performing Organ. Report No.	3. Publication Date October 1987

4. TITLE AND SUBTITLE

Computer Science and Technology:
Guide on Data Entity Naming Conventions

PB88-128799/AS

5. AUTHOR(S)

Judith J. Newton

6. PERFORMING ORGANIZATION (If joint or other than NBS, see instructions)

NATIONAL BUREAU OF STANDARDS
DEPARTMENT OF COMMERCE
~~WASHINGTON, D.C. 20234~~
Gaithersburg, MD 20899

7. Contract/Grant No.

8. Type of Report & Period Covered

Final

9. SPONSORING ORGANIZATION NAME AND COMPLETE ADDRESS (Street, City, State, ZIP)

Same as item 6.

NBS Category No.
NBS-160

10. SUPPLEMENTARY NOTES

Library of Congress Catalog Card Number: 87-619867

Also Available from
GPO as SN003-003-02818-5

☐ Document describes a computer program; SF-185, FIPS Software Summary, is attached.

11. ABSTRACT (A 200-word or less factual summary of most significant information. If document includes a significant

A coherent set of naming conventions for data entities is crucial to the central management of data. Name content and format must be designed to maximize the information content and relationship to the logical structure of the data. This report discusses the development and administration of naming conventions for data entities. The Information Resource Dictionary System (IRDS) meta-name schema provides a framework for name development. This report describes a methodology for deriving a relationship between an entity's dictionary names and details a method for structuring the format and content of entity names which maximizes opportunities for analysis and sharing of data.

12. KEY WORDS (Six to twelve entries; alphabetical order; capitalize only proper names; and separate key words by semicolons)

data administration; database; data dictionary system; data engineering; data entity naming conventions; data management; data standards; IRDS

13. AVAILABILITY

[X] Unlimited
☐ For Official Distribution. Do Not Release to NTIS
[X] Order From Superintendent of Documents, U.S. Government Printing Office, Washington, D.C. 20402.
☐ Order From National Technical Information Service (NTIS), Springfield, VA. 22161

14. NO. OF PRINTED PAGES

57

15. Price

TABLE OF CONTENTS

LIST OF FIGURES v

ACKNOWLEDGEMENTS vi

1. INTRODUCTION 1
   1.1. THE UTILITY OF NAMING CONVENTIONS 1
   1.2. INTENDED AUDIENCE FOR THIS GUIDE 1
   1.3. HOW THIS GUIDE SHOULD BE USED 1
   1.4. SOME TERMS DEFINED 2
   1.5. WHAT'S IN A NAMING CONVENTION? 3
   1.6. ADVANTAGES OF NAMING CONVENTIONS 5
   1.7. THE NATURE OF NAMES 7
   1.8. GUIDING PRINCIPLES FOR RULE DERIVATION 7

2. ORGANIZATIONAL FACTORS IN NAME DEVELOPMENT 9
   2.1. THE ROLE OF THE DATA ADMINISTRATOR 9
   2.2. MANAGEMENT SUPPORT 9
   2.3. MANAGEMENT STYLE 9
   2.4. FUNCTION 10
   2.5. SCOPE 10
   2.6. NON-TRADITIONAL COMPUTING 12

3. ADMINISTRATION OF NAMING CONVENTIONS 14
   3.1. TASKS OF THE DATA ADMINISTRATOR 14
      3.1.1. CONVENTIONS AND STANDARDS 14
      3.1.2. ADMINISTRATIVE PROCEDURES 14
   3.2. TOOLS 15
      3.2.1. THE DATA DICTIONARY 15
      3.2.2. OTHER TOOLS 16

4. DATA ARCHITECTURE 20
   4.1. DATA TYPES 20
   4.2. THE IRDS NAME STRUCTURE 20
   4.3. CONCEPTUAL STRUCTURE OF DATA 21
   4.4. NAMES AND LOGICAL STRUCTURE 22
   4.5. NAMES AND CLASSIFICATION 22
   4.6. NAMES AND DATA RATIONALIZATION 23
   4.7. CONVENTIONS FOR NON-ELEMENT ENTITIES 23

5. NAME FORMATION 24
   5.1. ACCESS NAMES 24
      5.1.1. ACCESS NAME CONTENT 24
      5.1.2. ACCESS NAME FORMAT 24
      5.1.3. ALTERNATIVE FORMATS 26
   5.2. DESCRIPTIVE NAMES 26
      5.2.1. DESCRIPTIVE NAME CONTENT 26
      5.2.2. DESCRIPTIVE NAME FORMAT 26
   5.3. RELATIONSHIP OF ACCESS AND DESCRIPTIVE NAMES . . . 27
   5.4. SPECIFYING FORMATS FOR DESCRIPTIVE NAMES 27

5.5. ALTERNATE NAMES	28
5.6. PHYSICAL IMPLEMENTATION	28
6. CONCLUSIONS	30
REFERENCES	31
APPENDIX A: CONSISTENT EXAMPLE	A-1
APPENDIX B: RULES FOR NAMING CONVENTIONS	B-1
APPENDIX C: GLOSSARY	C-1

LIST OF FIGURES

1: Content/Format Interrelationship........................4
2: Entity and Attribute Mapping............................6
3: Synonyms v. Same Names.................................11
4: Prototyping..13
5: Thesaurus Example.....................................19
6: IRDS Meta-Names.......................................20
7: Strategic Data Planning...............................22
8: Horizontal and Vertical Integration...................25
9: Schematic of Logical Groupings and Relationships.......A-2

ACKNOWLEDGEMENTS

The author wishes to acknowledge the assistance of the following in the preparation of this guide:

For background material on naming conventions:

    Rene Fecteau - Veterans Administration
    Steven Scheibe

    Dan Landon - Puget Sound Naval Shipyard

    Wm. "Mack" Leftwich - Controller of the Currency (formerly of USPS)

    Susan McFarland - IRS
    Ileane Solomon

    Chuck Spasaro - USPS
    Al Henderson
    Larry Willets

    Rae Thompson -Smithsonian Institution

For review and helpful criticism of this and earlier versions of this guide:

    Ellen Levin - Freddie Mac

    Ron Shelby - American Management Systems, Inc.

    Len Gallagher - National Bureau of Standards
    Alan Goldfine
    David Jefferson
    Frankie Spielman

GUIDE ON DATA ENTITY NAMING CONVENTIONS

Judith Newton

A coherent set of naming conventions for data entities is crucial to the central management of data. Name content and format must be designed to maximize the information content and relationship to the logical structure of the data. This report discusses the development and administration of naming conventions for data entities. The Information Resource Dictionary System (IRDS) meta-name schema provides a framework for name development. This report describes a methodology for deriving a relationship between an entity's dictionary names and details a method for structuring the format and content of entity names which maximizes opportunities for analysis and sharing of data.

Key words: data administration; database; data dictionary system; data engineering; data entity naming conventions; data management; data standards; IRDS.

1. INTRODUCTION

1.1. THE UTILITY OF NAMING CONVENTIONS

The mission of data administration is the management of data as a corporate resource [DURR85]. While all corporate data should be available to the appropriate users, a centralized section of the organization must be responsible for the oversight of data resources. The data administrator provides the strategy and tools to accomplish goals such as the reduction of costs of data collection, storage, and usage, and the reduction of errors in process and data design.

One of the strategies which the data administrator can use to further both the management and distribution of data is the adoption of a set of data entity naming conventions (DENC's). Used in conjunction with logical database design, naming conventions can provide greater efficiency of data handling, a cost savings in reduced computer time, and reduced confusion among both staff and management.

1.2. INTENDED AUDIENCE FOR THIS GUIDE

This guide is intended primarily for data administration staff concerned with establishing and administering naming conventions in their organizations, and also for database administrators concerned with establishing naming rules within their areas of responsibility. Although the sections dealing with the Federal Information Processing Standard (FIPS) Information Resource Dictionary System (IRDS) will be of most concern to the Federal community, the bulk of this guide is designed for general use.

1.3. HOW THIS GUIDE SHOULD BE USED

This document is intended to provide guidance on the establishment of a set of naming conventions for use within a government agency or business organization. It does not recommend a set of naming conventions to be applied to any organization regardless of the data to be addressed. Rather, it contains a set of factors which should be considered when developing and administering naming conventions. All examples contained within this document are examples, and not recommendations.

References in the text to the conceptual business model and logical data model are deliberately generic and non-specific; it is not the purpose of this guide to describe or discuss these aspects of information management, nor to distinguish among the many methodologies available.

Data names should reflect the content and relationships of the data which they identify. Each organization's body of data is unique [KONI81]. Each government agency will have information needs which differ from other government agencies. In addition, management style will dictate the mode of promulgation and enforcement. Therefore, this guide can only suggest ways to structure names which best serve the needs of each organization. The form of the names must be determined by the corporate data administration function in cooperation with the developers and users of the data.

Data entity naming conventions do not exist in a vacuum [SPIE86]. The data administrator must establish and maintain other conventions and standards, such as those related to data definition [FIPS126, FIPS127], life-cycle management [FIPS101 GOLD82], data security [FIPS73, FIPS113], data integrity [FIPS88], and quality assurance [FIFE77].

1.4. SOME TERMS DEFINED

Throughout this document, references to _data entity_, _data element_, _data entity name_, _data element name_, and _element name_ will be found. Although definitions of these terms appear in the Glossary (Appendix C), some clarification at this point will avoid confusion.

   o   _Data entity_ and _data entity name_ refer to a class of objects of concern to the organization, about which information is kept on a computerized system. Examples of data entities are systems, databases, files, reports, and data elements. They should not be confused with business entities which result from a conceptual business model and represent a higher level of conceptualization. Data entities represent the crucial component of efficient data sharing throughout an organization. Other possible types of entities include process, relationship and attribute entities. While naming conventions can be applied to these entities, the importance of standardized names is greatest for data entities. The names of these other entities are not propagated over the entire organization, nor are they used for data sharing as data entity names are.

Because many organizations store information using more than one retrieval method, the application of naming standards may be deliberately avoided. This is rationalized by the seeming hopelessness of reconciling all the differing name constraints (most often size constraints) of the systems used. The problem is resolved by using one overall <u>corporate name</u> combined with several <u>alternate names</u>. In this guide, data entity name is synonymous with corporate name unless otherwise specified.

    o   <u>Data element</u>, <u>data element name</u> and <u>element name</u> refer to a subset of data entities. Data elements represent the smallest logical division of information. Usually, there will be many more data elements than other types of data entities in an organization's database. Their names also tend to be more diverse and require more analysis than other data entities.

In addition, the terms <u>data entity naming conventions</u> and <u>naming conventions</u> are synonymous. The term <u>data element naming conventions</u> is used to refer specifically to data elements.

1.5. WHAT'S IN A NAMING CONVENTION?

There are two areas of concern when naming, or any aspect of language, is discussed: <u>content</u> and <u>format</u>. Any set of naming conventions should address both areas in terms of the organization's data.

    o   **Content** involves the essential meaning or significance of the words chosen for the name. It may be equated to the function of the semantic portion of language; that is, the assigned meanings of words.

    <u>Information content</u> is a term used to describe the amount of knowledge conveyed to the observer. This term may be seen as composed of two parts:

       - <u>Discrete content</u> describes the amount of information which may be derived about the subject entity by perusal of the data entity name.

       - <u>Relational content</u> describes the amount of information an observer may derive about other entities by perusal of the subject name.

    The degree to which either discrete or relational content should be maximized is an important decision for the data administrator.

    o   **Format** concerns the size, shape and general plan of organization or arrangement of the words in the name. It may be thought of as the syntactical portion of language; the ways in which words are put together. Many factors must be considered when the formats of names are decided upon, such as length, character set, word form, word order, abbreviations and acronyms, and connectors and modifiers. When formatting names, two levels of relationships must be considered:

- <u>Micro structure</u> concerns the arrangement and relationships of components of a name.

- <u>Macro structure</u> concerns relationships of names to other names and to the logical data structure.

Establishing rules for the formatting of names may not at first seem as important as establishing rules for content. Certainly it is tempting to establish formatting rules more quickly than content rules, for they may appear to be more straightforward than content rules. However, one area should not be given any less consideration during planning and administration than the other. Formatting rules play an important part in maintaining consistency of the data. Also, format and content are interrelated; the more thought is given to balancing the <u>information content axis</u>, the more care must be taken concerning the format areas of micro and macro structure. In Figure 1, a point x_1 on the content axis indicates a name with content weighted towards the discrete end of the axis. The corresponding point y_1 represents a name format weighted equally towards the micro structure end of the format axis. The value pair (x_1, y_1) represents that point on the <u>ideal balance line</u> where the best-formed name is found.

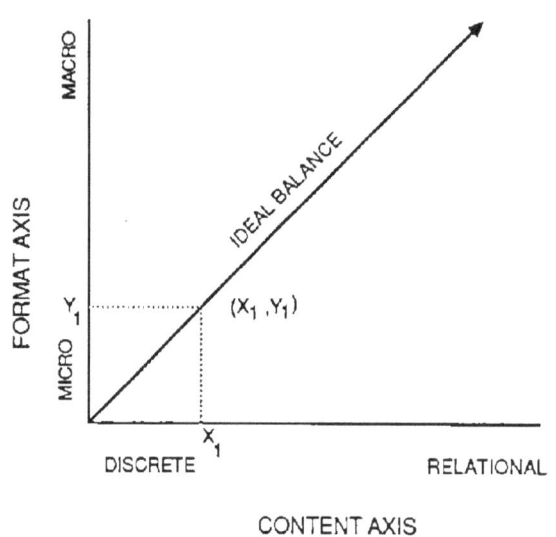

Figure 1: Content/Format Interrelationship:
Information Content Axis

As an example, consider one of the sets of names described in Appendix A: the logical grouping CONTRACTOR.

Information Content: CONTR-NAME is composed of two words: CONTR (an abbreviation of CONTRACTOR) and NAME. The information immediately conveyed to the reader is that "this is the name of a contractor the organization has (or may have or has had) dealings with." This is the discrete content. The relational content in this case is only implicit; that is, the reader must infer that since the name of the contractor is recorded in the data base, other information about this subject is recorded there as well. A search for other entities containing the string CONTR will reveal that there are at least three other data elements and one other record associated with contracting. Somewhere in this body of data will be the information needed to resolve the ambiguity of the discrete content.

Format: Regarding micro-structure, the arrangement of the words in CONTR-NAME and the use of the hyphen as separator make the meaning of the data element name easy to understand. It is very close to natural language order. The macro-structure is implicit in the abbreviation of CONTRACTOR to CONTR. As this name would be shorter than the (assumed) length limit even if not abbreviated, the abbreviation must have been used to maintain consistency with other, longer names in the same grouping. This is revealed by perusal of CONTR-CONTACT-NAME.

This discussion deliberately does not consider the additional meaning given to the names by the prime/class word organization described later in this guide, or the relationship between access and descriptive names.

1.6. ADVANTAGES OF NAMING CONVENTIONS

With the development of the data administration function within the organization, centralized oversight of data becomes possible. Among the advantages of this activity is the reduction of redundant data through consolidation of synonymous and overlapping data elements. This can be achieved by the application of rules which lead to the creation of consistent names.

In many organizations, development and maintenance of a conceptual business model (CBM) is a major activity of the data administration staff. Many of the tools used in the development of the business model also assist in the derivation of a logical data model [FONG85, MART82]. Characteristics of names which reflect this logical model provide mapping between the naming structure and the data's logical structure.

Many methodologies for deriving a logical data model use the E-R Approach for their underlying structure [CHEN77, NAVA86, ROSS87]. This guide assumes this basic organization both for the logical data model and in the structure of data dictionary entries for data entities (metadata). Use of the E-R Approach for data modeling results in a set of entities with related attributes. The data entities referred to in this guide will in most cases map to the attributes of E-R entities. In Figure 2, the E-R entity "Employee" has attributes which include "name" and "address." These objects become the logical group named EMPLOYEE which contains the data entities EMP-NAME and EMP-ADDR.

Naming conventions also assist in the classification of data. Organization of data into categories such as codes, numerical, dates, etc. can be explicitly expressed in the name. This facilitates many kinds of analysis.

Lastly, names which represent data entities in a clear and descriptive way are greatly preferred to those given without thought to data sharing or future use by others. Nondescriptive or misleading names have no place in the organization for which total integration of data is a significant goal.

E - R Model

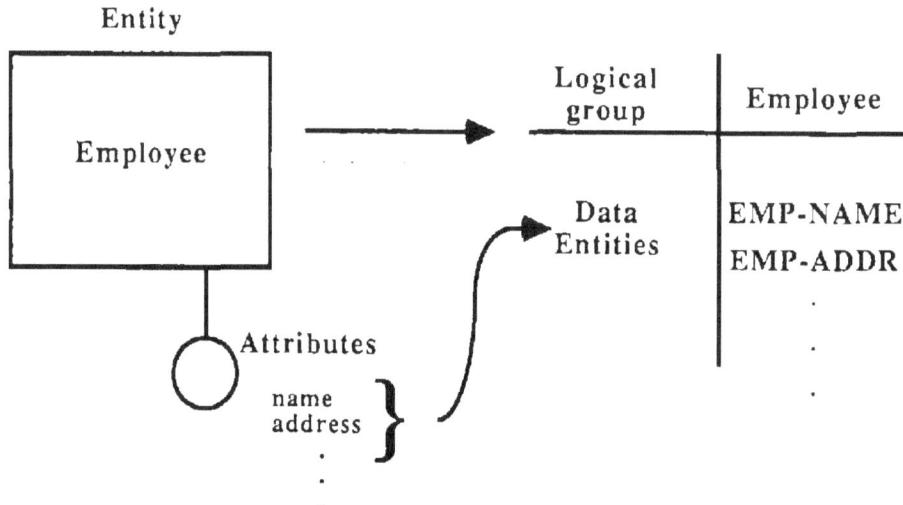

Figure 2: Entity and Attribute Mapping

1.7. THE NATURE OF NAMES

Data entity names differ from common names, or nouns, in the same ways that programming languages differ from natural languages; both are operating under <u>constraints</u> which impose a need for strict control over their structure, content and usage.

Data entity names are constrained by:

    o   Hardware - Although less of an influence since the advent of third-generation languages, hardware can still exert indirect constraint, for example, in limiting the choice of tools such as data dictionary systems.

    o   Software - Operating systems limit sort field size, dictate the maximum size and content of file names, and often impose constraints on other entity names, such as program, module, and report names. Programming languages impose limits on format and content of data element names used within the programs. COBOL has a thirty-character limit, while C, a more recently-developed language gaining in popularity, restricts significance to the first eight characters of data names.

    o   Database Management Systems, Fourth-Generation Languages, etc. - These newer products can be surprisingly old-fashioned about length and format restrictions. One well-known database product allows only ten characters for record field names. Data dictionary systems also constrain both allowed length and structure of data entity names, and often dictate the schema by which the data is described.

    o   The Human Factor - Resistance to change by users can be a challenge to data administration. Those who have been using established names for years often see no reason to change. It is natural for users to want the greatest possible expressivity for new names but also brevity for keying and (human) memory recall. Also, policies from higher levels of management can impose restrictions on naming format or content.

1.8. GUIDING PRINCIPLES FOR RULE DERIVATION

The application of a set of consistent data entity naming conventions represents an opportunity to express the logical structure of an organization's data. As a tool for both data administrator and end user, the inherent advantages can be irresistible. Yet naming conventions can represent a temptation as well as an opportunity: the temptation to become locked too rigidly into a scheme which may not be appropriate to the data being named. Naming conventions involve a tradeoff between structure and flexibility. The more strongly specified the structure of the names and the relationship of those names to the structure of the data, the greater the possibility for sacrificing legibility and comprehensibility of the names.

To achieve the best possible mapping between the logical data model and data entity names, some guiding principles should be followed when naming conventions are developed [NEWT84].

> o   <u>Clarity</u> - Names are as clear and English-like as possible. Ideally, they are immediately obvious to the casual user. Although it is seldom possible to achieve this goal for all names, the principle must always be kept in mind.
>
> o   <u>Brevity within Uniqueness</u> - Names are as short as possible while still retaining meaning and uniqueness within the database. Any conflicts between this and the principle of clarity are resolved in favor of the latter.
>
> o   <u>Conformance to Rules of Syntax</u> - Each name is in the proper format. Waivers, if granted, are used sparingly. The degree of specificity of format rules will drive the frequency of waiver requests.
>
> o   <u>Context-Freedom</u> - Each entity is considered discretely from all others. The name references the logical structure but is as independent as possible from the physical structure of the data and from other data entities. For example, the name of a data element collected from a form does not contain the name or number of the form. Relationships and other data documented in the data dictionary entry for an entity are not part of the name.

2. ORGANIZATIONAL FACTORS IN NAME DEVELOPMENT

2.1. THE ROLE OF THE DATA ADMINISTRATOR

The function and placement of data administration within an organization are still in a process of evolution. Some data administrators are viewed as an interface between the database administrators (DBA's) and the end users, while others interact with users independently. Data administration can be a separate function from database administration or contained within an integrated information systems department. Often, a data administrator's first job is to convince management that the function is necessary; even more often, the function must be defined before the job may begin [JAME87].

2.2. MANAGEMENT SUPPORT

Management support can be crucial to the success of data administration. An unlucky data administrator may be placed in the position which database administrators and proponents of data dictionaries used to find themselves in the past, with an upper-level management indifferent to data administration support. In that case, neither naming conventions nor any other function assumed by data administration has much prospect for success.

Many data administrators develop charters [DURR85]. These documents define the functions of the data administration staff and the scope of its activities. They also map the relationships between data administration and DBA, management, and end users. A charter can be a powerful tool for legitimization of the data administration function.

2.3. MANAGEMENT STYLE

The culture of the organization will determine the management style of the data administrator. When upper management fosters an autocratic image, data administration will tend to issue directives without much interaction beforehand. If consensus is the order of the day, however, a more cooperative atmosphere will result. Most organizations fall somewhere between these two extremes [FONG86].

2.4. FUNCTION

The major function of data administration is to oversee the logical design and application of all the data of concern to the organization. Traditionally, implementation of a logical data model has been the occupation of the database administrator. This often led to an emphasis on physical design and to a tendency to get lost in the forest while trying to avoid bumping into trees. Data administration instead concentrates on seeing the forest as a whole and understanding its ecology.

2.5. SCOPE

Ideally, a data administration function is placed to oversee the information management of the entire organization; this is how the greatest long-term benefit is derived. When the decision is made to design and implement a conceptual business model, a data administration function should be established concurrently. Then the logical data model can be derived from the conceptual business model and implemented forthwith.

If data administration is divided among organizational units, good communication must be maintained. Sometimes, units of an organization will establish the function independently. This does nothing to further organization-wide data sharing.

The task of applying naming convention rules should first be applied to new systems. If old systems are being redesigned, they should be considered candidates for revised names as well. Old systems in place, however, should not be disturbed, except for the inclusion of their data entities in the organizational data dictionary for documentation. If these names do not conform to the new rules, they should be included as synonyms for the new dictionary entries. The advantages of synonym reduction must be weighed against the impact of change to each system.

Synonyms - two or more occurrences of the same data entity with differing names - cause confusion among those sharing data across systems. Even instances of synonyms being used within one system have been discovered. Although an ideal of data administration is to reduce synonyms to the greatest extent possible, the documentation of all synonyms in the data dictionary can be established as a workable goal. One way to resolve the issue of which name should be 'the' name is to establish a corporate name for the entity which is different from all the known synonymous names; the latter are then listed as <u>aliases</u> of the corporate name. Aliases, which appear in the dictionary entry for the entity with relationships to the context in which they occur in logical or physical databases, are considered <u>controlled synonyms</u>.

For instance, there may be three records in a system which contain information about an employee. Figure 3 shows how three different names for the same data element (EMP-NAME) may be consolidated. Refer to Appendix A for a data dictionary entry description of EMP-NAME which documents aliases.

Homonyms must also be controlled. These are two different data entities which share the same name. They hold much greater potential for causing problems than uncontrolled synonyms. Where synonyms are considered a passive problem (resulting in redundant data storage and manipulation), homonym use actively creates errors in results caused by using the wrong data entity - a case of mistaken identity. The best policy for the data administrator is to require uniqueness of every entry name in the dictionary. This will force resolution of homonym occurrences.

<u>EMP-RCRD</u>   <u>TRAIN-RCRD</u>
EMP-NAME   TRAINEE-NAME

<u>CONTRACT-POC-GROUP</u>
NAME-OF-POC
PHONE-NO-OF-POC

3A: Three Different Names for the Same Data Element

<u>EMP-RCRD</u>   <u>TRAIN-RCRD</u>
EMP-NAME   EMP-NAME

<u>CONTRACT-POC-GROUP</u>
EMP-NAME
EMP-PHONE-NMBR

3B: Using the Same Name Facilitates Analysis

Figure 3 : Synonyms v. Same Names

2.6. NON-TRADITIONAL COMPUTING

Data administrators work towards the standardization of data management within an organization. Some new developments are making this task more difficult than before. Changes in the traditional data life cycle are affecting the entire "data climate" of many organizations.

> o Microcomputer use makes it very easy for end users to create and maintain data outside the organization's "standards umbrella." But sometimes those who develop software on micros find their former stand-alone systems being converted to organization-wide or multi-site distribution. Others download data from mainframe databases and then want to reload after manipulating the data. Requiring any data being uploaded into the common system from a micro to meet all the DENC and other data standards of the organization is the most important rule for the data administrator to promulgate. Other auditing and quality assurance standards may also be imposed [STAN87].

> o Prototyping - the rapid development of subsets of large systems using new software tools developed for this purpose - may lead to use of nonstandard data entity names in the interest of speedy development. But creating names "on the fly" leads to delays during the implementation of the production system when conversion to the organization's DENC standards becomes necessary (Figure 4). Using DENC's at the initiation of the prototyping effort enables developers to avoid the "retrofit pit" and leads to more efficient use of prototyping in the long run [GRAY86, FISH87].

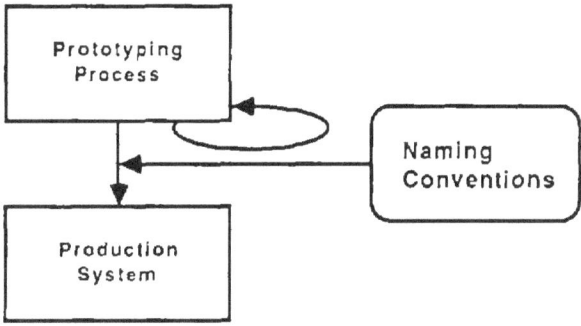

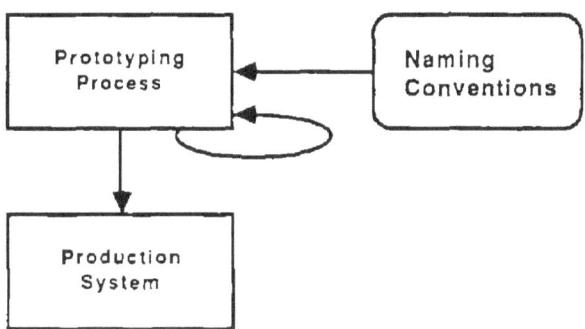

Figure 4: Prototyping

3. ADMINISTRATION OF NAMING CONVENTIONS

3.1. TASKS OF THE DATA ADMINISTRATOR

The data administrator is the driving force behind the promulgation and enforcement of data naming standards in an organization. To this end, the data administrator uses a variety of tools and procedures.

3.1.1. CONVENTIONS AND STANDARDS

A <u>convention</u> is a rule or algorithm which, when applied to a body of data, produces a <u>standard</u> result. A set of naming conventions is a tool for the data administrator's use to assist in maintaining consistent data entity naming standards and for the data administrator and the end user to use to produce standard data entity names.

The development of naming conventions is a cooperative effort by the data administrator and the end users. The goal is the establishment of flexible, easy to use conventions which will result in a consistent set of data entity names.

The DENC's are embodied in a set of rules which are distributed throughout the organization. The names which users derive from the application of these rules are approved by the data administrator and entered in the data dictionary. An example set of rules is presented in Appendix B. Clearly, enforcement authority must be granted to data administration by high-level management; this can be established in the data administration charter.

3.1.2. ADMINISTRATIVE PROCEDURES

There are five administrative activities associated with naming data entities:

    o Establish data ownership. The relationship between users and their data varies among organizations. Usually, the originator of a data entity is viewed as the "owner" of that entity and has sole control over its creation (including naming), modification, and updating, but not deletion. The owner is responsible for notifying other users when the entity is altered (usually with the assistance of the data administrator) and must approve any changes proposed by other users. The term "owner" is falling into disfavor in some organizations, to be replaced by "custodian" or "primary user." This reflects a concept that is becoming more widely accepted: that the organization is the real owner of all the data.

o   Establish a relationship between the users and data administration. The functions of each in regard to naming data entities must be clear. A procedure for the resolution of conflicting names for identical data entities should be included.

o   Establish a procedure for approval of standard names. This process involves four steps: originate, approval by data administration, enter in data dictionary or other repository, and promulgate. Each of these may be either paper or automated.

o Write the DENC manual. The document detailing the rules for DENC derivation may be incorporated into a general administrative procedures manual. Copies of the forms associated with submission, approval and dissemination of standard names should be included, together with a description of the process. This document is written by the data administrator with input from users.

o   Promulgate the rules. Be sure everyone involved with data handling throughout the organization is aware of the DENC standards and knows how to apply the rules. Formal classes may be needed to convince users of the advantages involved.

3.2. TOOLS

3.2.1. THE DATA DICTIONARY SYSTEM

The data dictionary system is the primary tool for the administration of naming conventions. All names and the relationships among them can be recorded and, if the dictionary system is an active one, controlled. Many dictionary systems are closely linked to database management systems and can record only the data used in that DBMS; the data administrator should have a dictionary system which is capable of managing all corporate (shared) data in an organization.

Data dictionary systems provide a schema, or structure, for describing data entities [LEFK83]. Names are among the characteristics specified in the schema. Often, use of a certain dictionary system will determine the structure and relationships among the names of a single data entity, and among those of different entities.

Name format is also determined by the dictionary system. A maximum length restriction is usually imposed on names. The format in which the examples in dictionary system manuals are presented may become the implicit format for users' names, in addition to the explicitly stated rules, such as "no spaces allowed between words."

But the advantages of a dictionary system far outweigh any constraints it may impose. It gives a data administrator the means to manage the different versions of data entities being used, often under differing names, by the various divisions of the organization or by software packages with diverging name constraints.

Appendix A shows examples of the output of a dictionary system report on entity entries. It is structured in IRDS format (see Section 4.2, The IRDS Name Structure). One corporate name has been assigned to each entry with the differing names required by different software packages and programming languages included as aliases. The data administrator should ascertain that only one alias per package or language is being used for corporate data for each entity.

Other reports or responses to ad hoc queries can be used to facilitate data analysis. For instance, a list of entity names and associated aliases gives the data administrator an overview of all names used by entities across systems, retrieval methods and departments. By properly qualifying the query, this list could be used to:

- o Track and control synonym use.

- o Uncover and eliminate use of nonstandard names.

- o Trace usage of a single data entity over time.

- o Identify all users with access to a certain data entity across one or more user views.

3.2.2. OTHER TOOLS

Although a data dictionary can provide a good base for the management of data entity names, some specialized tools can make the data administrator's job easier. They fall into two categories: automated packages that work with the data dictionary system, and data lists (either automated or manual) developed by data administration and tailored to the DENC's of the organization.

Automated packages may be either purchased or "home-grown." Most of those available for purchase interface with the data dictionary to provide name analysis and control functions superior to those of the dictionary alone [ADPA, COMP, GLOB, WATE]. Among them are:

o   Synonym identification. Most packages provide this service by producing a report listing all similar names or a permuted word list (Keyword in context [kwic] list).

o   Generation of standard names by automatic application of the rules specified by the data administrator.

o   Identification of names that do not fit the organization's format and content standards.

o   Automatic application of abbreviation rules.

o   Operating system and programming language audit trails.

o   Replacement of non-standard names in program source code and generation of an input file with new names for the data dictionary.

o   Quality assurance and impact analysis reports.

o   Generation of graphic structure diagrams and/or bubble charts.

Computer Aided Software Engineering (CASE) systems are emerging as a new way to integrate the processes of data modeling, logical and physical database design and implementation, and data management with a coherent set of tools designed to work together [HULI87]. They may also incorporate a methodology for conceptual business model development [MART86, TATA, TIPP84, TIPP85, TSII84].

Although these systems vary in content, emphasis on a particular area of the range described above, and degree of integration, they all contain a dictionary facility as a centerpiece to coordinate the functions of their components. Often, they glorify the dictionary with a name such as "repository" or "encyclopedia" to reflect the enhanced role it plays in the control of system information. This central use of the dictionary facilitates the use of DENC's early in the design process and enhances the ability to maintain naming consistency throughout the application process. Among the other principle components of CASE systems are:

o   Diagramming Tools

o   Syntax Verifier

o   Prototyping Tools

o   Code Generators

o   Life-Cycle Methodology Tools

o   Project Management Tools

The data administrator needs to control the format of the words composing data entity names. This is accomplished through the construction and maintenance of lists of <u>abbreviations</u>, <u>acronyms</u>, and <u>allowed words</u>. These lists are made available to users either on-line or on paper.

o   Abbreviation list: there is only one approved abbreviation for each word. A methodology for creating abbreviations is included in the DENC manual to allow users to create them when needed (see Appendix B for a sample methodology). All abbreviations are then approved by the data administrator and entered onto the abbreviation list.

o   Acronym list: A list of the acronyms in common use by the organization. This includes both common business terms (e.g., COB for close of business) and proper names usually used in acronym form (SECNAV for Secretary of the Navy). Only one acronym is allowed per term.

o   Allowed words: Prime and class words are restricted to those appearing on the relevant list. A very rigorous naming environment may also restrict modifiers. (These concepts will be discussed in Sections 4.4 and 4.5.)

A <u>glossary</u> is a useful document for aiding users in the proper choice of words for a name. Definitions of important words used in names simplify the choices and reduce the chance of proliferation of synonymous names. Appendix B provides a sample extract from a data entity word glossary.

One tool which combines the first three functions described above is the <u>thesaurus</u> [GENE78]. It is a known and accepted format for organizing and controlling terminology in the library and information science disciplines. A thesaurus consists of a word list structured in a standard way to express a hierarchy of terms. For example, in the GAO Thesaurus five types of cross-references are used:

o   USE: this word is the preferred term over any synonym, near-synonym, or word-form variant.

o UF: 'used-for' or 'used in lieu of' - this term should not be used. There is a synonym or near-synonym which is preferred.

o BT: broader term - A term of greater specificity than the entry.

o NT: narrower term - A term of lesser specificity than the entry.

o RT: - a related term which does not fall into a hierarchy.

For an excerpt of thesaurus material, see Figure 5. The entry under NAME provides the following information: NAME should be used in preference to TITLE and ID; the broader term is DESIGNATION while the narrower term is LABEL; and a related term is CODE. The entries for TITLE and ID specify that the term NAME should be used instead.

A thesaurus can be automated to interface with a data dictionary. [CORN87]. It can capture the definitions and relationships of business terms, which is the traditional role of a thesaurus; it can also use these terms to edit and validate the data entity names in the dictionary. The business terms are stored on a traditional thesaurus, or 'keyword' file. This file is then used to derive others for the major content words of a name and abbreviations/acronyms. Each name can be checked against the appropriate files for rule conformance.

<u>NAME</u>

UF Title
     ID
BT Designation
NT Label
RT Code

<u>TITLE</u>

USE Name

<u>ID</u>

USE Name

Figure 5: Thesaurus Example

4. DATA ARCHITECTURE

4.1. DATA TYPES

Most data fall into one of two type categories: business or technical (scientific). Some technical or scientific data is less amenable to standardization than business data (test result items will be in numerical representation) or the names will have already been standardized by a scientific discipline (chemical or natural order names). Business names will vary depending on the focus of the business or mission of the government organization. Before the application of any set of naming conventions, the body of data relevant to an organization should be analyzed to determine which convention most readily fits the majority of entities.

4.2. THE IRDS NAME STRUCTURE

Before development of DENC's can commence there must exist a structure and nomenclature for the names themselves. The quantity and variety of names for each entity, and their relationships to each other and to other entity attributes, must be decided. So must the names of the names (meta-names).

The Information Resource Dictionary System (IRDS) is an emerging national, Federal and international standard. The structure and terminology for assigning names to data entities described there will be used in this guide (Figure 6) [GOLD85, ANSI86].

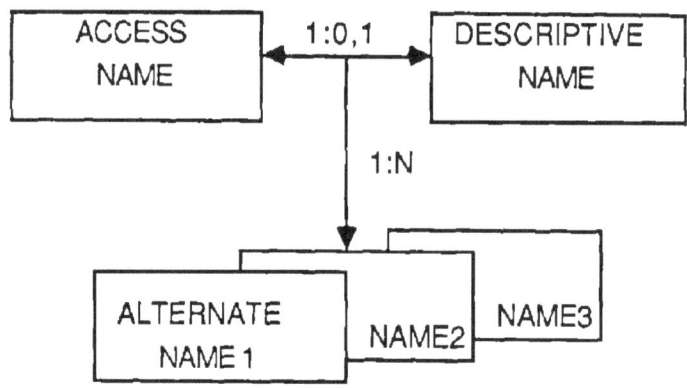

Figure 6: IRDS Meta-Names

o   <u>Access Name</u> is the primary, unique identifier of the data entity. In most organizations, this name will be terse. Each dictionary entry has only one access name.

o   <u>Descriptive Name</u> is normally longer than but functionally the same as the access name. Each dictionary entry has none or one descriptive name.

o   <u>Alternate Name(s)</u> are functionally attributes of the entity, are not unique, and serve as aliases. There may be many alternate names for any one dictionary entry.

The advantage of having two names, both of which can be used to refer to a single data entity (especially useful in the case of data elements), can be derived when users at differing levels of familiarity with the database use the dictionary. The access name is used by those who have more experience with the data and would be impatient with typing a long name for ad hoc queries. Other, more casual users would be more comfortable with the descriptive name, which is more English-like and may provide more information than the access name.

In addition, the access name is freed of the burden of being the only identification for an entity. The temptation to 'load' information into the name is reduced, and the need to abbreviate is also reduced. Thus, the access name becomes less cryptic than it might have been.

A natural process of forming access names is described below. Descriptive names are derived from them.

4.3. CONCEPTUAL STRUCTURE OF DATA

<u>Strategic data planning</u> involves the formulation of a conceptual business model (CBM) and the derivation of a <u>conceptual data model</u> (CDM) (Figure 7). From the CDM two other objects are developed [GRIE82]:

o   the <u>external model</u>, also known as the user's view or logical data model (LDM); and

o   the <u>internal model</u>, the physical model from which the physical implementation is produced.

The conceptual structure of data unique to each organization should be the driving force behind the DENC structure. This organization is expressed in the LDM, which contains <u>logical data entities</u>.

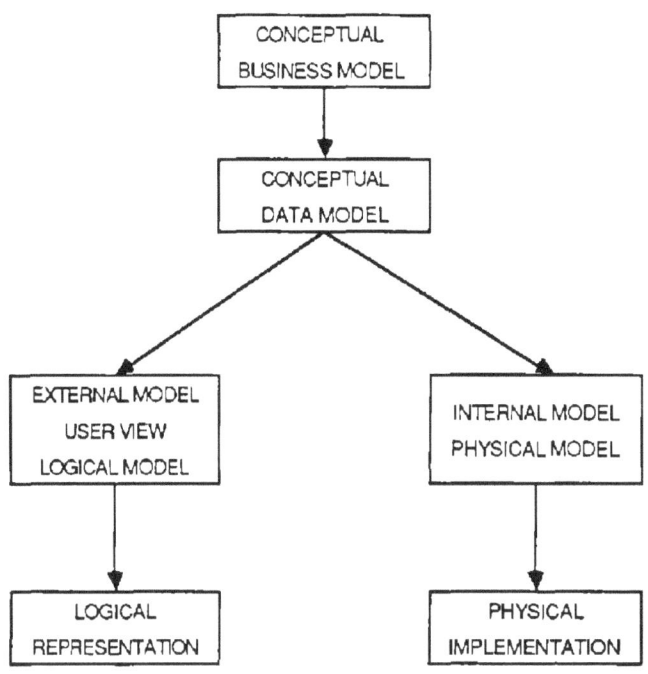

Figure 7: Strategic Data Planning

4.4. NAMES AND LOGICAL STRUCTURE

Mapping the logical structure of the data in the data entity name represents an opportunity to impose organization upon a potentially chaotic situation. By including in the data entity name a form of the name of the logical database to which the data entity belongs, a relationship is established which conveys valuable information to the user. Thus the information content of the name can be increased. As this part of the name is its most significant piece of information, this word is known as the Prime Word. For example, all data elements relating to employees of an organization may carry EMPLOYEE as a prime word.

4.5. NAMES AND CLASSIFICATION

Most, if not all, data entities can be classified into specific categories of information. A classification may be thought of as the answer to the question, "What is it?" The answer might be "a code," "a date," "a file," or "a line of text." This information about what type of data composes this data entity is of second-most significance to the user. This word is known as the Class Word. It may be condensed into a three- or four-letter code and restricted to only those codes approved by the data administrator. For instance, all names of objects (persons, places, or things) may contain the class word NAME.

For data entities other than data elements, the class word signifies the <u>entity-type</u>. The IRDS, together with most data dictionary systems, categorizes data entities into groups such as <u>data elements</u>, <u>files</u>, <u>databases</u>, <u>reports</u>, etc. In most cases, the software then automatically establishes relationships between data entities based on these entity-types.

Once the prime and class words have been determined, modifiers added to the name ensure uniqueness from other data entities.

4.6. NAMES AND DATA RATIONALIZATION

Each data element in the database should express one idea or represent one object. Good naming conventions can help data rationalization by making it difficult to apply a well-formed name (one that follows the conventions) to an unrationalized element. For instance, the rule of allowing one and only one prime word and one and only one class word per name limits the concepts which can be described. If an element seems to require two of either word, that is a clue that the element should be decomposed into two or more elements.

This technique is particularly useful when old systems are being redesigned and their data elements are renamed in the new DENC format. Those which cannot be fit into the format should be analyzed and restructured when necessary.

4.7. CONVENTIONS FOR NON-ELEMENT ENTITIES

Entity-types such as records, files, and databases follow the same format as data elements. The class words usually need only be identifiers of the entity-type, e.g., EMP-RCRD or EMPLOYEE-RECORD. (Only logical records are discussed here and in most other places in this guide. For physical record application, see Section 5.6, Physical Implementation.) If many formats for entity names are in use, which will be the case in multiple-software environments, one corporate name should be used in the organization standard format with all non-standard names listed as alternate names. See Appendix A for examples. Care should be taken to avoid listing entities with different physical implementation formats under the same corporate name.

5. NAME FORMATION

The method for forming access and descriptive names described below is one possible way to structure standard names. It may not be the best way for every organization, but is an example of how to approach the subject and the topics of concern.

5.1. ACCESS NAMES

5.1.1. ACCESS NAME CONTENT

The access name contains one prime word and one class word together with modifiers to identify, describe and make the name unique. As the IRDS requires uniqueness of all data entity names within the dictionary, qualification is not considered here.

The use of prime and class words allow both vertical and horizontal analysis of the data (Figure 8). Prime words group data logically (horizontally), giving the analyst clues about other closely related entities (see Section 1.5, What's In a Naming Convention?). Class words relate a particular entity to others of the same category (vertically).

Restricting the membership in the sets of prime and class words to certain predetermined words allows greater rigor in the control of synonyms. A list of class word definitions will prevent the use of EMP-NO when only EMP-ID-CODE should be used.

5.1.2. ACCESS NAME FORMAT

The access name should be terse. While remaining as readable as possible, it must have a length limitation which makes it usable in ad hoc retrievals; between 30 and 40 characters, depending on the needs of the organization.

The arrangement of words in the access name is decided by the data administrator and the users. Function words - articles, conjunctions, and prepositions - are not allowed except to establish uniqueness.

The placement of prime words, class words and modifiers within a name is most important for maximum readability. Of course, the data administrator will want to take advantage of the mapping to logical and classificational categories which the content of the names provide by using tools designed for data analysis, but software currently available makes rigid placement of certain words for automated processing unnecessary. The decision on placement becomes one which depends on the human users of the information. Readability seems maximized when the format:

> PRIME WORD : MODIFIER(S) : CLASS WORD

is used. This is the format used for all examples in this guide. Colons are used above for separation emphasis only.

Other format issues must also be resolved and rules formulated to address them. These include a list of allowable characters, allowed abbreviations and acronyms together with rules for abbreviation formation (see Section 3.2.2, Other Tools), connectors between words, word form (voice and tense), and capitalization. Appendix B presents examples of these rules. In the examples presented in this guide, use of hyphens as word separators is arbitrary.

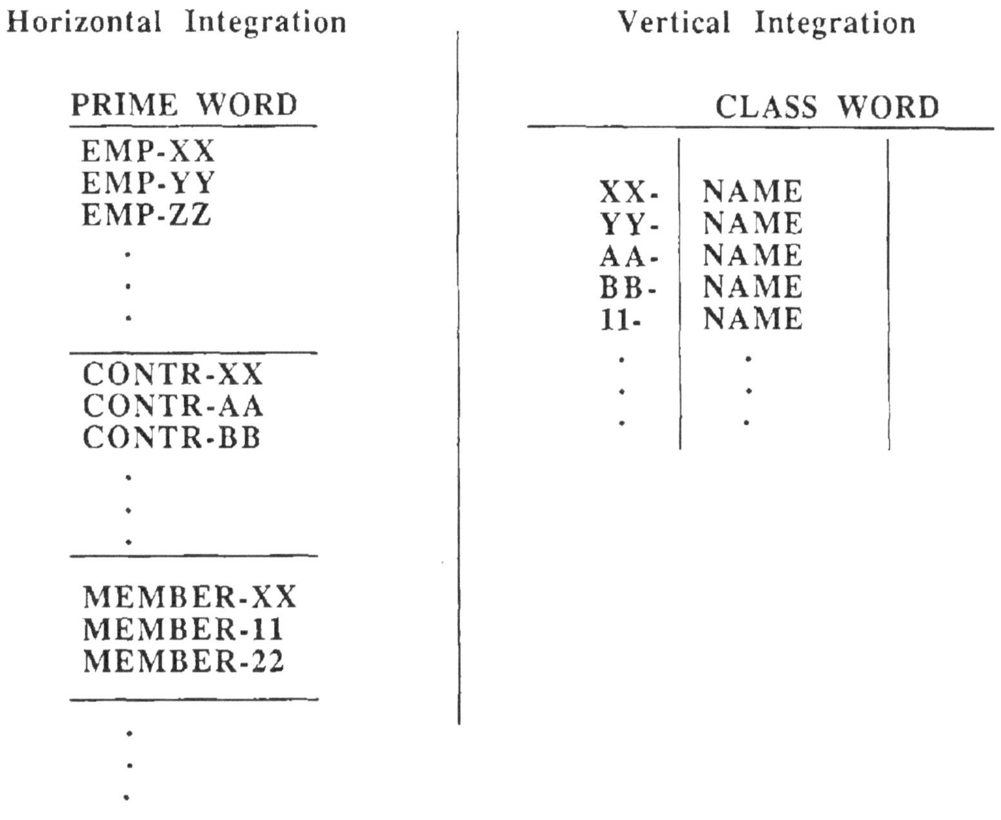

Horizontal Integration	Vertical Integration
PRIME WORD	**CLASS WORD**
EMP-XX	XX- NAME
EMP-YY	YY- NAME
EMP-ZZ	AA- NAME
.	BB- NAME
.	11- NAME
.	. .
CONTR-XX	. .
CONTR-AA	. .
CONTR-BB	
.	
.	
.	
MEMBER-XX	
MEMBER-11	
MEMBER-22	
.	
.	
.	

Figure 8: Horizontal and Vertical Integration

5.1.3. ALTERNATIVE FORMATS

Modifiers need not be grouped in the middle of the name. An example of an alternative format is:

> MODIFIER(S) : PRIME WORD : MODIFIER(S) : CLASS WORD

Although this format has the benefit of linking the modifiers with either the prime or class word, the accompanying disadvantage is the possible confusion of the prime word with its modifier(s). For example, in the element name TRAINING-COURSE-NMBR, the prime word may be either TRAINING or COURSE.

Another alternative format is:

> CLASS WORD : PRIME WORD : MODIFIER(S)

The data administrator determines the format after analysis of the organization's data and consultation with users.

5.2. DESCRIPTIVE NAMES

5.2.1. DESCRIPTIVE NAME CONTENT

The descriptive name is a natural language phrase which contains at least the prime and class words of the access name. Most of the modifiers of the access name are found in the descriptive name as well, though this is not an absolute requirement. In addition, as many additional connectors and modifiers are added as is necessary to produce a readable and grammatical phrase.

The brevity principle should not, however, be ignored when forming descriptive names. A size limit will in most cases still apply. There should be no danger of confusing a name with a definition. Nothing which appears in the rest of the dictionary entry for the entity should be included in the name. Such information might be the origin of an entity (e.g., a form), or representational data (format or length).

5.2.2. DESCRIPTIVE NAME FORMAT

The format of the descriptive name is that of an English noun phrase. Ideally, it contains spaces between words and allows flexibility in word order so that the most readable name possible is developed.

The relationship between a descriptive name and an access name is illuminated by an example. The access name CONTR-CONTACT-NAME is developed into a descriptive name by expanding the contraction CONTR to CONTRACTOR, dropping the hyphens, rearranging the word order and adding a connector. It becomes CONTACT NAME OF CONTRACTOR. It is good policy to avoid altering the form of the prime word; use ...NAME OF CONTRACTOR instead of CONTRACTOR'S NAME.

5.3. RELATIONSHIP OF ACCESS AND DESCRIPTIVE NAMES

Although the access and descriptive names of the same data entity contain the same prime and class words, this is not a foolproof way to identify one when presented with the other. It is possible to have two names within the same logical group both be codes or dates. In this case, modifiers become especially important and should be given careful attention when deriving one name from the other.

5.4. SPECIFYING FORMATS FOR DESCRIPTIVE NAMES

Some rules for the formation of descriptive names:

  o A descriptive name must conform to the rules of English grammar and usage.

  o It must be a single, discrete phrase - not a sentence or sentence fragment.

  o It must not contain an active verb; a name such as DATE MEMBER JOINS is inappropriate. JOIN DATE OF MEMBER is preferred.

These are some examples of descriptive names, based on the access names in Appendix A.

Access Name	Descriptive Name
EMP-RCRD	Employee Record
EMP-NAME	Name of Employee
EMP-ADDR	Address of Employee
EMP-SOC-SEC-NMBR	Social Security Number of Employee
EMP-ID-CODE	Identification Code of Employee

5.5. ALTERNATE NAMES

Alternate names consist of aliases for the corporate name of the data entity (see Section 2.5, Scope). It may be possible for those alternate names which are variations because of the constraints imposed by different hardware or software systems to conform to a suitably restricted form of naming convention. An effort should be made to restrict the alternate names to one from each different system. Other forms of alternate name, such as report headings, are less amenable to standardization.

Appendix A shows that the alternate names for EMP-RCRD are: EMPLOYEE-REC, used by FOCUS; EMP_REC, used by a C language program; and E-REC, a COBOL file name.

5.6. PHYSICAL IMPLEMENTATION

At physical design time, most of the logical data entities become physical data entities with no change in the names. Some of them may have to be decomposed or combined in the interest of efficiency or normalization.

Other entities must be created, for instance, to represent relationships between files in a relational database. Fields known as "foreign keys" exist only to provide this link. The data administrator may want to create a special class word KEY to identify these entities.

Most data dictionary systems are oriented towards documenting the physical representations of data entities, with little attention paid to the logical data entities from which the former are derived. The prime word may be the only indication in a dictionary entry of the relationship to logical groupings of the logical data model.

The documentation of logical entities in the dictionary may be accomplished in a variety of ways. Some of them are:

> o A separate copy of the data dictionary for logical data entities. The physical data entities are listed as alternate names in the logical dictionary entries. The physical dictionary description contains the corresponding logical dictionary entity name.

> o Logical data entity names used as corporate names with physical names listed as alternate names in the same dictionary entry. All physical implementations sharing the entry have the same format. If they differ, they have their own entries.

o The use of extensibility to create new entity-types for logical entities, with associated relationship- and attribute-types. This would permit the listing of all associated physical entities in the entry for each logical entity, including representation information for each.

6. CONCLUSIONS

A robust set of naming conventions is a valuable tool for mapping logical data structure and classification to data entities. In addition, it can reduce to a rational, orderly process the confusion often associated with assigning names.

When data entity names reflect the structure and content of the database, the advantages provided to data analysis are manifold. Consistent names also facilitate the benefits of data sharing to users in different parts of the organization.

Application of the principles of clarity, brevity, rule conformance and context-freedom assist in developing conventions which will produce names in close conformance to the organization's standards. Like most design activities, the effort expended in advance of their application will pay off over the life of the enterprise.

REFERENCES

[ADPA]   Adpac Computing Languages Corporation, *PM/SS Functions and Features*, Adpac, San Francisco, CA, n.d.

[ANSI86]   ANSI X3H4, *(Draft Proposed) American National Standard Information Resource Dictionary System, Parts 1-6*, American National Standards Institute, New York, 1986.

[CHEN77]   Chen, P. P., "The Entity-Relationship Model-Toward a Unified View of Data," *ACM Transactions on Database Systems*, Vol. 1, No. 1, March 1976.

[COMP]   Composer Technology Corporation, "Data Dictionary Composer," CTC, Santa Clara, CA, n.d.

[CORN87]   Cornwell, Marilyn, "Thesaurus: Management Control Framework for the Dictionary," *Data Dictionary Symposium Proceedings*, Atlantic City, NJ, June 1987.

[DURR85]   Durell, William, *Data Administration: A Practical Guide to Successful Data Management*, McGraw-Hill, New York, 1985.

[FIFE77]   Fife, Dennis W., *Computer Software Management: A Primer for Project Management and Quality Control*, NBS Special Publication 500-11, National Bureau of Standards, Gaithersburg, MD, July 1977.

[FIPS73]   Federal Information Processing Standards Publication 73, *Guidelines for Security of Computer Applications*, National Bureau of Standards, U.S. Department of Commerce, June 30, 1980.

[FIPS88]   Federal Information Processing Standards Publication 88, *Guideline on Integrity Assurance and Control in Database Administration*, National Bureau of Standards, U.S. Department of Commerce, August 14, 1981.

[FIPS101]   Federal Information Processing Standards Publication 101, *Guideline for Lifecycle Validation, Verification, and Testing of Computer Software*, National Bureau of Standards, U.S. Department of Commerce, June 6, 1983.

[FIPS113]   Federal Information Processing Standards Publication 113, *Computer Data Authentication*, National Bureau of Standards, U.S. Department of Commerce, May 30, 1985.

[FIPS126]   Federal Information Processing Standards Publication 126, *Database Language NDL*, National Bureau of Standards, U.S. Department of Commerce, March 10, 1987.

[FIPS127]   Federal Information Processing Standards Publication 127, *Database Language SQL*, National Bureau of Standards, U.S. Department of Commerce, March 10, 1987.

[FISH87]   Fisher, Gary E., *Application Software Prototyping and Fourth Generation Languages*, NBS Special Publication 500-148, National Bureau of Standards, Gaithersburg, MD, May 1987.

[FONG85]   Fong, E. N., Henderson, M. W., Jefferson, D. K., and Sullivan, J. M., *Guide on Logical Database Design*, NBS Special Publication 500-122, National Bureau of Standards, Gaithersburg, MD, February 1985.

[FONG86]   Fong, E., and Goldfine, A., eds., *Data Base Directions: Information Resource Management-Making It Work*, NBS Special Publication 500-139, National Bureau of Standards, Gaithersburg MD, June 1986.

[GENE78]   General Accounting Office, *General Accounting Office Thesaurus*, Washington DC, November 1978.

[GLOB]   Global Software, Inc., "A Rose by Any Other Name [$NAME]," Global Software, Duxbury, MA, n.d.

[GOLD82]   Goldfine, Alan H., ed., *Data Base Directions Information Resource Management - Strategies and Tools*, NBS Special Publication 500-92, National Bureau of Standards, Gaithersburg, MD, September 1982.

[GOLD85]   Goldfine, Alan, and Konig, Patricia, *A Technical Overview of the Information Resource Dictionary System*, NBSIR 85-3164, National Bureau of Standards, Gaithersburg, MD, April 1985.

[GRAY86]   Gray, Martha Mulford, *Guide to the Selection and Use of Fourth Generation Languages*, NBS Special

Publication 500-143, National Bureau of Standards, Gaithersburg, MD, September 1986.

[GRIE84] Griethuysen, J. J. van (Ed.), *Concepts and Terminology for the Conceptual Schema and the Information Base*, ISO/TC 97/SC 21/Document N197. (Also ISO/TC 97/SC 5/WG3, Document N695, March 1982.)

[HULI87] Huling, Jim, "Key Elements of CASE Kits: Prototyping, Code Generators," *Computerworld*, Vol. XXI No. 16, April 20, 1987.

[JAME87] James, David, "Selling Data Administration for the 80's and Beyond," *Guide 67*, Anaheim, CA, March, 1987.

[KONI81] Konig, Patricia A., and Newton, Judith J., *Federal Requirements for a Federal Information Processing Standard Data Dictionary System*, NBSIR 81-2354, National Bureau of Standards, Gaithersburg, MD, September 1981.

[LEFK83] Lefkovits, Henry C., Sibley, Edgar H., and Lefkovits, Sandra L., *Information Resource/Data Dictionary Systems*, QED Information Sciences, Inc., Wellesley, MA, 1983.

[MART82] Martin, James, *Strategic Data-Planning Methodologies*, Prentice-Hall, Inc., Englewood Cliffs, NJ, 1982.

[MART86] Martin, James, and Hershey, E.A. III, *Information Engineering: A Management White Paper*, Knowledge-Ware, Ann Arbor, MI, 1986.

[NAVA86] Navathe, Shamkant, Elmasri, Ramez, and Larson, James, "Integrating User Views in Database Design," *Computer*, IEEE, January 1986.

[NEWT84] Newton, Judith J., "Observations on Data Element Naming Conventions," *Proceedings, Trends and Applications 1984*, IEEE Computer Society Press, May 1984.

[ROSS87] Ross, Ronald G., *Entity Modeling: Techniques and Application*, Database Research Group, Inc., Boston, MA, 1987.

[SPIE86] Spielman, Frankie E., ed., *Data Administration Workshop Proceedings*, NBSIR 86-3324, National

Bureau of Standards, Gaithersburg, MD, February 1986.

[STAN87] Stang, David J., "Micro Manager's Responsibilities Seem Unending," <u>Government Computer News</u>, May 22, 1987.

[TATA] TATA Consultancy Services, "Advanced Data Dictionary (ADDICT)," TATA, New York, NY, n.d.

[TIPP84] Technology Information Products Corporation, <u>Data Catalog 2 System Overview</u>, TIP, Bulington, MA, 1984.

[TIPP85] Technology Information Products Corporation, <u>TIP Plan Starter Kit</u>, TIP, Burlington, MA, 1985.

[TSII84] TSI International Ltd., <u>FACETS System Overview</u>, TIP, Burlington, MA, 1984.

[WATE] Waterfield Company, The, <u>Data Expediter Technical Overview</u>, Waterfield Co., Fairfax, VA, n.d.

APPENDIX A: CONSISTENT EXAMPLE

These data entity names are found throughout this guide as examples. The logical groupings in which the data entities appear represent functional divisions of the data. Different divisions within the organization such as the payroll, personnel and training departments share the information grouped under the heading PERSONS. Other logical groups, such as TRAINING, CONTRACTS, and CONTRIBUTIONS, are tied to those in PERSONS through relationships. See Figure 9 for a schematic representation of the logical groupings.

These logical data entities may be decomposed into groups of data elements or become record, file or database names at physical design time.

PERSONS

EMPLOYEE	CONTRACTOR	MEMBER
EMP-RCRD	CONTR-RCRD	MEMBER-RCRD
EMP-NAME	CONTR-NAME	MEMBER-NAME
EMP-ADDR	CONTR-ADDR	MEMBER-ADDR
EMP-SOC-SEC-NMBR	CONTR-ID-CODE	MEMBER-ID-CODE
EMP-ID-CODE	CONTR-CONTACT-NAME	MEMBER-JOIN-DATE
EMP-START-DATE		

TRAINING	CONTRACT	CONTRIBUTION
TRAIN-RCRD	CONTRACT-RCRD	CONTRIB-RCRD
TRAIN-COURSE-DESC	CONTRACT-NMBR	CONTRIB-AMNT
TRAIN-COURSE-NMBR	CONTRACT-DATE	CONTRIB-DATE
TRAIN-DATE	.	CONTRIB-DESC
.	.	.
.	.	.
.	.	.

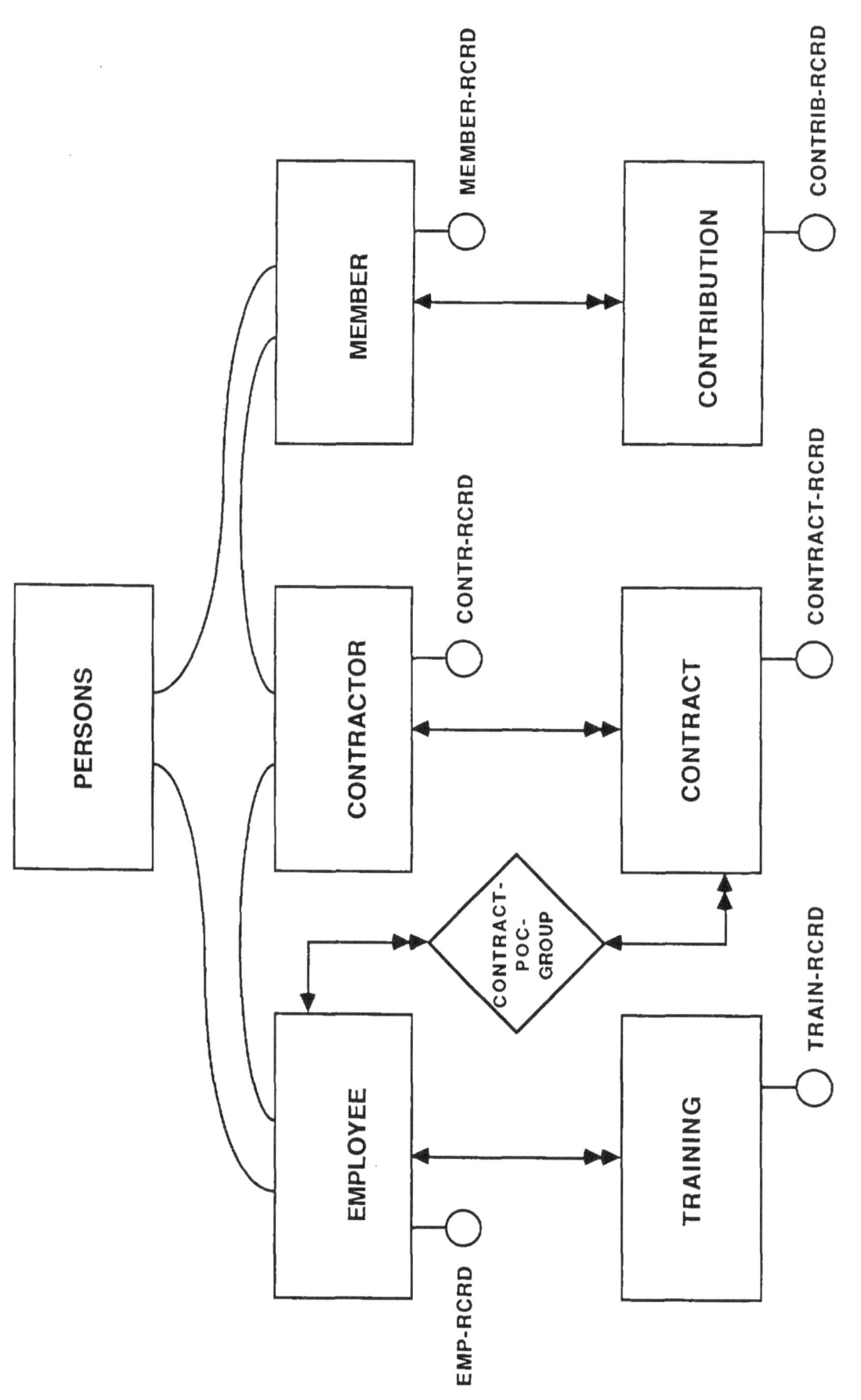

Figure 9: Schematic of Logical Groupings and Relationships

A-2

EXAMPLES OF DATA DICTIONARY ENTRIES

These samples of entries as they appear in a full listing report are typical of the output of the IRDS. In particular, they were generated in the format adopted by the IRDS Prototype project at the Institute of Computer Sciences and Technology (ICST). The access, descriptive, and alternate names are shown, as well as the attributes and relationships involving the entries.

****

ENTITY = EMP-RCRD   (This is the access name.)
DESCRIPTIVE_NAME = EMPLOYEE RECORD
ENTITY_TYPE = RECORD

ATTRIBUTES

ADDED_BY = NEWTON

DESCRIPTION = EMPLOYEE RECORD is used as a central repository of personal information about employees.

NUMBER_OF_TIMES_MODIFIED = 0

GROUP ATTRIBUTES

DATE_TIME_ADDED

SYSTEM_DATE = 20870707
SYSTEM_TIME = 083745

IDENTIFICATION_NAMES

ALTERNATE_NAME = EMPLOYEE-REC
ALTERNATE_NAME_CONTEXT = FOCUS

ALTERNATE_NAME = EMP_REC
ALTERNATE_NAME_CONTEXT = C LANGUAGE

ALTERNATE_NAME = E-REC
ALTERNATE_NAME_CONTEXT = COBOL WORKING FILE

RELATIONSHIPS

RECORD EMP-RCRD CONTAINS ELEMENT EMP-NAME
RECORD EMP-RCRD CONTAINS ELEMENT EMP-ADDR
RECORD EMP-RCRD CONTAINS ELEMENT EMP-SOC-SEC-NMBR
RECORD EMP-RCRD CONTAINS ELEMENT EMP-ID-CODE
RECORD EMP-RCRD CONTAINS ELEMENT EMP-START-DATE

SYSTEM PERSONS PROCESSES RECORD EMP-RCRD
****

```
*****************************************************************
```

ENTITY = EMP-NAME
DESCRIPTIVE_NAME = EMPLOYEE NAME
ENTITY_TYPE = ELEMENT

ATTRIBUTES

ADDED_BY = NEWTON

DESCRIPTION = The full name of an employee, including the first, last, and middle.

LENGTH = 70 CHARS

COMMENTS = E-NAME has its own entry in this dictionary.

NUMBER_OF_TIMES_MODIFIED = 2

GROUP ATTRIBUTES

DATE_TIME_ADDED

SYSTEM_DATE = 20870707
SYSTEM_TIME = 083745

IDENTIFICATION_NAMES

ALTERNATE_NAME = E-NAME
ALTERNATE_NAME_CONTEXT = COBOL WORKING FILE

ALTERNATE_NAME = NAME OF EMPLOYEE
ALTERNATE_NAME_CONTEXT = EMP SYSTEM REPORT HEADER

RELATIONSHIPS

ELEMENT EMP-NAME STANDARD FOR ELEMENT E-NAME

RECORD EMP-RCRD CONTAINS ELEMENT EMP-NAME

```
*****************************************************************
```

```
***************************************************************

ENTITY = E-NAME
ENTITY_TYPE = ELEMENT

                        ATTRIBUTES

ADDED_BY = NEWTON

DESCRIPTION = This element is the COBOL representation of EMP-
NAME.  It is a group element composed of three subelements: L-
NAME, F-NAME, and M-NAME in that order.
      L-NAME = 30 characters
      F-NAME = 20 characters
      M-NAME = 20 characters

LENGTH = 70 CHARS

NUMBER_OF_TIMES_MODIFIED = 1

                     GROUP ATTRIBUTES

        DATE_TIME_ADDED

     SYSTEM_DATE = 20870707
     SYSTEM_TIME = 083745

                       RELATIONSHIPS

RECORD E-REC CONTAINS ELEMENT E-NAME

ELEMENT E-NAME STANDARD TO ELEMENT EMP-NAME

***************************************************************
```

APPENDIX B: RULES FOR NAMING CONVENTIONS

This is a sample set of rules for DENC's that would appear in a standards manual. Some of the rules have alternatives; these will appear in brackets. All are examples to follow when establishing DENC's in an organization, to be adapted to the individual enterprise.

I. Purpose and Scope of Naming Standards

The use of standard data entity names promotes the accessibility of data across systems by insuring consistency of data names and data definitions. It also allows users access to documentation about names as used throughout the organization.

The rules documented below assure a standardized set of names with consistent format and content. Corporate names are used as dictionary entry names and must conform to the rules. These names may be logical names which represent multiple physical implementations of entities; this is represented by physical names appearing as alternate names (synonyms) in the dictionary.

Two forms of name are used: access name and descriptive name. Each will be described below.

II. Procedure for Naming Entities

The structure of a standard access name is shown below:

> PRIME WORD : MODIFIER(S) : CLASS WORD

A. Access Name Content

Prime words will be selected from the list of prime words and should reflect the logical grouping of the entity as referenced in the chart showing groups and relationships of data entities within this organization. [Note: this chart should represent the logical data model of the organization.]

Class words will be selected from the list of class words and represent the class or category of the data.

Modifiers will be added as needed to describe the entity and make it unique within the dictionary.

One and only one prime word, and one and only one class word, are required. Modifiers may be added until the length limit is reached.

If a prime or class word is needed which does not appear on the appropriate list, consult the data administrator for possible inclusion.

Abbreviations are allowed, but must come from the list of abbreviations, or be formulated according to the methodology described below and approved by the data administrator for inclusion on the list. [Abbreviations may be made mandatory for words over a certain length.]

The policy on acronyms is the same as that for abbreviations.

B. Access Name Format

    1. Nouns are used in singular form only. Verbs are in the present tense.

    2. Words are separated by hyphens. No other special characters are allowed.

    3. All words in the name are in capital letters.

    4. Names are limited to 35 characters in length.

C. Descriptive Name Content and Format

Descriptive names are expanded forms of the access names. They are free-format. They must contain the prime and class words found in the access name, and may contain additional modifiers and connectors such as "of" and "for." They do not contain abbreviations. Words are separated by spaces. The descriptive name sounds as much like an English phrase as possible.

The descriptive name is limited to 80 characters in length.

III. Prime and Class Word Lists

A. Prime Words

Extracts of the prime and class word lists are presented in

glossary format. The full form of the logical grouping name, an adapted form which is used as the prime word, and a definition are shown. The class words have been abbreviated to four-letter codes.

LOGICAL GROUP NAME	PRIME WORD	DEFINITION
EMPLOYEE	EMP	A person employed by this organization.
CONTRACT	CONTRACT	A contract between this organization and another.
CONTRACTOR	CONTR	The organization with which this organization has contract(s).
CONTRIBUTION	CONTRIB	A donation made by a member.
MEMBER	MEMBER	A person or organization which pays dues to this organization.
TRAINING	TRAIN	Training in job expertise given to an employee of this organization.

B. Class Words

Class words define the category to which an entity belongs. Data elements have many possible class words, other entities only one. The number of class words for data elements varies with the needs of the organization, but the set below is a workable minimum. The class words for non-element entities conform to the Basic Functional Schema of the IRDS.

CLASS WORD	CODE	DEFINITION
DOCUMENT	DCMT	A human readable data collection.
FILE	FILE	An instance of an organization's data collection.

MODULE	MODL	An automated process; either a logical subdivision of PROGRAM or an independent process called by a PROGRAM.
PROGRAM	PROG	An instance of an automated process.
RECORD	RCRD	An instance of logically associated data.
SYSTEM	SYST	A collection of processes and data.
USER	USER	Individuals or organizational components.

ELEMENT:

ADDRESS	ADDR	The designation of a place of residence or receipt of mail.
AMOUNT	AMNT	Monetary quantity.
AVERAGE	AVRG	Numeric value representing an arithmetic mean.
COUNT	CONT	Non-monetary numeric value arrived at by counting.
CODE	CODE	A system of valid symbols which substitute for longer values.
DATE	DATE	Calendar date.
NAME	NAME	A designation for an object.
NUMBER	NMBR	A number associated with an object.
QUANTITY	QNTY	Non-monetary numeric value not arrived at by counting.
RATE	RATE	A quantity or amount considered in relation to another quantity or amount (e.g., miles/gallon).

TEXT	TEXT	An unformatted descriptive field.
TIME	TIME	Time of day or duration.

IV. Modifiers

A modifier list is used when rigorous control of all words in the access name is desired. Allowed modifiers, their approved abbreviations, and a short definition are included. A sample is shown below.

MODIFIER	ABBREVIAT'N	DEFINITION
CONTACT	CNTCT	A person or organization designated as monitor (e.g., of a contract).
COURSE	CRSE	A course of instruction.
IDENTIFIER	ID	Number or code which identifies an object.
JOIN	JOIN	Associate; members may join this organization by paying dues.
SECURITY	SEC	Used in the phrase SOCIAL SECURITY NUMBER (SOC-SEC-NMBR) only.
SOCIAL	SOC	Used in the phrase SOCIAL SECURITY NUMBER (SOC-SEC-NMBR) only.

V. Abbreviation Methodology

A. Eliminate vowels right to left to form a meaningful abbreviation. Never delete the first letter of the word.

    Example: CNTCT for contact

B. Use a short form of the word if it is easily recognized.

    Example: ID for Identifier.

C. Do <u>not</u> use an abbreviation that is a word in its own right.

    Example: ALTER for Alternate.

D. Do <u>not</u> use hyphens, slashes, or other special characters.

E. Do <u>not</u> use an abbreviation that reproduces a prime or class word.

    Example: CONT (the code for COUNT) for Continued.

F. A word can have only one abbreviation. A particular abbreviation can be used for only one word.

VI. Acronyms

An extract of a sample acronym list appears below.

ACRONYM	EXPANDED TERM
AWOL	Absent Without Leave
ATF	Alcohol, Tobacco and Firearms
AKA	Also Known As
BPD	Bureau of the Public Debt
CPU	Central Processing Unit
COA	Change of Address
DP	Data Processing
POC	Point of Contact

. . . etc.

APPENDIX C - GLOSSARY

Access Name	In the IRDS, the most important identifier of an entity.
Alias	A controlled synonym of a data entity, one that is documented in a data dictionary or elsewhere.
Alternate Name	Another name for a data entity. Documented as an attribute in the IRDS.
Attribute Type	A characteristic of a data entity type.
Class Word	A word in the name of a data entity describing the category to which the data entity belongs, e.g., "file," "date," "name."
Classification	The process of breaking down a general group of entities into specific categories.
Computer Aided Software Engineering (CASE) tools	A group of tools designed to work together to integrate the process of data modeling and data management.
Conceptual Business Model (CBM)	A model of the business functions of the organization, including data flows and stores, and business processes.
Conceptual Data Model (CDM)	A model concentrating on and expanding the data aspects of the CBM.
Content	The essential meaning or significance of an object, such as a name.
Corporate Name	The name given to a data entity to represent that entity wherever it occurs in the information process of the organization; usually a logical name.
Data Administration (DA)	That function of the organization which oversees the management of data across all functions of the organization, and is responsible for central information planning and control.

Data Dictionary System (DDS)	A (usually automated) system for the documentation and control of data entities.
Data Element	An atomic unit of information.
Data Entity	An object of interest to the enterprise, usually tracked by an automated system.
Data Entity Naming Convention (DENC)	A rule which, when applied to a body of data entities along with other rules, results in a set of standardized names.
Descriptive Name	In the IRDS, an optional name, functionally the same as an access name but longer and more descriptive.
Discrete Content	The amount of information which may be derived about a data entity by perusal of the data entity name.
Entity (E-R Model)	This term is approximately equivalent to a logical data group as used in this guide.
E-R Model	A technique for modeling an organization's data, originally proposed by Peter Chen, involving the description of entities and the relationships between them.
External model	A model of the logical data structure which is independent of physical implementations. Also known as the user view or logical model.
Format	The size, shape and general plan of organization or arrangement of the components of an object, such as a name.
Function Words	Grammatical indicators such as prepositions, conjunctions, articles, and auxiliary verbs.
Homonym	A data entity with the same name as another data entity, but which differs from the latter in some essential characteristic.

Information Content	Describes the amount of knowledge conveyed to the observer upon perusal of an object, such as a name.
Internal Model	The model for a physical implementation of a data processing system. Also known as the physical model.
Information Resource Dictionary System (IRDS)	An emerging Federal, national and international standard for data dictionary system implementation. The basis for the data entity naming structure in this guide.
Logical Data Entity	A component of the logical data model which can be modeled in the data dictionary and may be the corporate name of the data entity.
Logical Data Model (LDM)	A model of the data stores and flows of the organization derived from the conceptual business model.
Macro Structure	The relationship of names to other names and to the logical data structure.
Metadata	Data about data; the names and attributes of data entities as stored in the data dictionary.
Meta-name data	Data about the name of a data entity; the type, amount and relationship of a single data entity's names.
Micro Structure	The arrangement and relationship of elements within a name.
Modifier	A word which helps define and render a name unique within the database, which is not the prime or class word.
Physical Data Entity	A data entity which is used in a physical implementation of the data model.
Prime Word	A word included in the name of a data entity which represents the logical data grouping (in the logical data model) to which it belongs; e.g., EMPLOYEE.

Relational Content	The amount of information an observer may derive about other entities by perusal of a data entity name.
Strategic Data Planning (SDP)	An activity of the data administrator designed to provide a comprehensive and controlled overview of the data resources of an organization.
Synonyms	Two or more occurrences of the same data entity under differing names.

NBS Technical Publications

Periodical

Journal of Research—The Journal of Research of the National Bureau of Standards reports NBS research and development in those disciplines of the physical and engineering sciences in which the Bureau is active. These include physics, chemistry, engineering, mathematics, and computer sciences. Papers cover a broad range of subjects, with major emphasis on measurement methodology and the basic technology underlying standardization. Also included from time to time are survey articles on topics closely related to the Bureau's technical and scientific programs. Issued six times a year.

Nonperiodicals

Monographs—Major contributions to the technical literature on various subjects related to the Bureau's scientific and technical activities.

Handbooks—Recommended codes of engineering and industrial practice (including safety codes) developed in cooperation with interested industries, professional organizations, and regulatory bodies.

Special Publications—Include proceedings of conferences sponsored by NBS, NBS annual reports, and other special publications appropriate to this grouping such as wall charts, pocket cards, and bibliographies.

Applied Mathematics Series—Mathematical tables, manuals, and studies of special interest to physicists, engineers, chemists, biologists, mathematicians, computer programmers, and others engaged in scientific and technical work.

National Standard Reference Data Series—Provides quantitative data on the physical and chemical properties of materials, compiled from the world's literature and critically evaluated. Developed under a worldwide program coordinated by NBS under the authority of the National Standard Data Act (Public Law 90-396).
NOTE: The Journal of Physical and Chemical Reference Data (JPCRD) is published quarterly for NBS by the American Chemical Society (ACS) and the American Institute of Physics (AIP). Subscriptions, reprints, and supplements are available from ACS, 1155 Sixteenth St., NW, Washington, DC 20056.

Building Science Series—Disseminates technical information developed at the Bureau on building materials, components, systems, and whole structures. The series presents research results, test methods, and performance criteria related to the structural and environmental functions and the durability and safety characteristics of building elements and systems.

Technical Notes—Studies or reports which are complete in themselves but restrictive in their treatment of a subject. Analogous to monographs but not so comprehensive in scope or definitive in treatment of the subject area. Often serve as a vehicle for final reports of work performed at NBS under the sponsorship of other government agencies.

Voluntary Product Standards—Developed under procedures published by the Department of Commerce in Part 10, Title 15, of the Code of Federal Regulations. The standards establish nationally recognized requirements for products, and provide all concerned interests with a basis for common understanding of the characteristics of the products. NBS administers this program as a supplement to the activities of the private sector standardizing organizations.

Consumer Information Series—Practical information, based on NBS research and experience, covering areas of interest to the consumer. Easily understandable language and illustrations provide useful background knowledge for shopping in today's technological marketplace.
Order the above NBS publications from: Superintendent of Documents, Government Printing Office, Washington, DC 20402.
Order the following NBS publications—FIPS and NBSIR's—from the National Technical Information Service, Springfield, VA 22161.

Federal Information Processing Standards Publications (FIPS PUB)—Publications in this series collectively constitute the Federal Information Processing Standards Register. The Register serves as the official source of information in the Federal Government regarding standards issued by NBS pursuant to the Federal Property and Administrative Services Act of 1949 as amended, Public Law 89-306 (79 Stat. 1127), and as implemented by Executive Order 11717 (38 FR 12315, dated May 11, 1973) and Part 6 of Title 15 CFR (Code of Federal Regulations).

NBS Interagency Reports (NBSIR)—A special series of interim or final reports on work performed by NBS for outside sponsors (both government and non-government). In general, initial distribution is handled by the sponsor; public distribution is by the National Technical Information Service, Springfield, VA 22161, in paper copy or microfiche form.

www.ingramcontent.com/pod-product-compliance
Lightning Source LLC
Chambersburg PA
CBHW081854170526
45167CB00007B/3014